A Challenge to the JOHNS HOP

ONE HUNDRED PROOFS

THAT

THE EARTH

IS

NOT A GLOBE.

Dedicated to RICHARD A. PROCTOR, Esq.
"The Greatest Astronomer of the Age."

By WM. CARPENTER,

*Referee for John Hampden, Esq., in the Celebrated Scientific Wager, in 1870;
Author of 'Common Sense' on Astronomy, (London, 1866;) Proctor's
Planet Earth; Wallace's Wonderful Water; The
Delusion of the Day, &c., &c.*

"UPRIGHT, DOWNRIGHT, STRAIGHTFORWARD."

BALTIMORE:
PRINTED AND PUBLISHED BY THE AUTHOR,
No. 71 Chew Street.
1885.

INDEX

1 The aeronaut sees for himself.
2 Standing water level.
3 'Surveyors' allowance.
4 Flow of Rivers — The Nile.
5 Lighthouses — Cape Hatteras.
6 The sea-shore. — " Coming up."
7 A trip down Chesapeake Bay.
8 The model globe useless.
9 The sailor's level charts.
10 The mariners' compass.
11 The southern circumference.
12 Circumnavigation of the Earth.
13 Meridians are straight lines.
14 Parallels of latitude — circles.
15 Sailing down and underneath.
16 Distance round the South.
17 Levelness required by man.
18 The "level" of the astronomers.
19 Half the globe is cut off, now!
20 No "up" or "down" in nature?
21 The "spherical loadstone."
22 No falsehoods wanted!
23 No proof of "rotundity."
24 A "most complete" failure.
25 The first Atlantic Cable.
26 Earth's "curvature."
27 Which end goes down?
28 A "hill of water."
29 Characteristics of a globe.
30 Horizon — level with the eye.
31 Much too small a globe.
32 Vanishing point of objects.
33 We are not "fastened on."
34 Our "antipodes." — a delusion.
35 Horizon a level line.
36 Chesapeake Bay by night.
37 Six months day and night.
38 The "Midnight Sun."
39 Sun moves round the Earth.
40 Suez Canal — 100 miles — level.
41 The "true level." — a curve.
42 Projectiles — firing east or west.
43 Bodies. thrown upwards.
44 Firing in opposite direction.
45 Astronomer Royal of England.
46 An utterly meaningless theory.
47 Professor Proctor's cylinder.
48 Proctor's false perspective.
49 Motion of the clouds,
50 Scriptural proof — a plane.
51 The " Standing Order."
52 More ice: in the south.
53 Sun's accelerated place, south.
54 Balloons'not left behind.
55 The Moon's beams are cold.
56 The Sun and Moon.
57 Not Earth's shadow at all.
58 Rotating and revolving.
59 Proctor's big mistake.
60 Sun's distance from Earth.
61 No true. "measuring-rod."
62 Sailing "round" a thing.
63 Telescopes—"hill of water."
64 The laws of optics — Glaish
65 "Dwelling" upon error.
66 Ptolemy's predictions.
67 Canal in China— 700 miles.
68 Mr, Lockyer's false logic.
69 Beggarly alternatives.
70 Mr. Lockyer's suppositions.
71 North Star seen from S. lat.
72 "Walls not parallel!"
73 Pendulum experiments.
74 "Delightful uncertainty."
75 Outrageous calculations.
76 J. R. Young's Navigation.
77 "Tumbling over."
78 Circumnavigation — south.
79 A disc — not a sphere.
80 Earth's "motion" unproven.
81 Moon's motion east to west.
82 All on the wrong track.
83 No meridianal "degrees.'
84 Depression of North Star.
85 Rivers flowing up-hill?
86 100 miles in five seconds.
87 Miserable makeshifts.
88 What holds the people down?
89 Luminous objects.
90 Practice against theory.
91 Unscientific classification.
92 G. B. Airy's "suppositions."
93 Astronomers give up theory.
94 School-room "proofs" false.
95 Pictorial proof — Earth a plane.
96 Laws of perspective ignored.
97 "Rational suppositions."
98 It is the star that moves.
99 Hair-splitting calculation.
100 How "time" is lost or gained.

ONE HUNDRED PROOFS

THAT

THE EARTH

IS

NOT A GLOBE.

Dedicated to R. A. PROCTOR, Esq.

BY WM. CARPENTER.

"UPRIGHT, DOWNRIGHT, STRAIGHTFORWARD."

[COPYRIGHT SECURED.]

BALTIMORE:
PRINTED AND PUBLISHED BY THE AUTHOR,
No. 71 Chew Street.
1885.

INTRODUCTION.

"PARALLAX," the Founder of the Zetetic Philosophy, is dead; and it now becomes the duty of those, especially, who knew him personally and who labored with him in the cause of Truth against Error, to begin, anew, the work which is left in their hands. Dr. Samuel B. Rowbotham finished his earthly labours, in England, the country of his birth, December 23, 1884, at the age of 89. He was, certainly, one of the most gifted of men: and though his labours as a public lecturer were confined within the limits of the British Islands his published work is known all over the world and is destined to live and be republished when books on the now popular system of philosophy will be considered in no other light than as bundles of waste paper. For several years did "Parallax" spread a knowledge of the facts which form the basis of his system without the slightest recognition from the newspaper press until, in January, 1849, the people were informed by the "Wilts Independent" that lectures had been delivered by "a gentleman adopting the name of 'Parallax,' to prove modern astronomy unreasonable and contradictory," that "great skill" was shown by the lecturer, and that he proved himself to be " thoroughly acquainted with the subject in all its bearings." Such was the beginning —the end will not be so easily described. The Truth will always find advocates—men who care not a snap of their fingers for the mere opinion of the world, whatever form it may take, whilst they know that they are the masters of the situation and that Reason is King! In 1867, "Parallax" was described as "a paragon of courtesy, good temper, and masterly skill in debate." The author of the following hastily-gotten-up pages is proud of having spent many a pleasant hour in the company of Samuel Birley Rowbotham.

A complete sketch of the "Zetetic Philosophy" is impossible in a small pamphlet; and many things necessarily remain unsaid which, perhaps, should have been touched upon, but which would to some extent have interfered with the plan laid down—the bringing together, in a concise form, "One Hundred Proofs that the Earth is not a Globe." Much may be gathered, indirectly, from the arguments in these pages, as to the real nature of the Earth on which we live and of the heavenly bodies which were created FOR US. The reader is requested to be patient in this matter and not expect a whole flood of light to burst in upon him at once, through the dense clouds of opposition and prejudice which hang all around. Old ideas have to be gotten rid of, by some people, before they can entertain the new; and this will especially be the case in the matter of the Sun, about which we are taught, by Mr. Proctor, as follows: "The globe of the Sun is so much larger than that of the Earth that no less than 1,250,000 globes as large as the Earth would be wanted to make up together a globe as large as the Sun.' Whereas, we know that, as it is demonstrated that the Sun moves round over the Earth, its size is proportionately less. We can then easily understand that Day and Night, and the Seasons are brought about by his daily circuits round in a course concentric with the North, diminishing in their extent to the end of June, and increasing until the end of December, the equatorial region being the area covered by the Sun's mean motion. If, then, these pages serve but to arouse the spirit of enquiry, the author will be satisfied. The right hand of fellowship in this good work is extended, in turn, to Mr. J. Lindgren, 90 South First Street, Brooklyn, E. D., N. Y., Mr. M. C. Flanders, lecturer, Kendall, Orleans County, N. Y., and to Mr. John Hampden, editor of "Parallax" (a new journal), Cosmos House, Balham, Surrey, England

ONE HUNDRED PROOFS
THAT
EARTH IS NOT A GLOBE.

IF man uses the senses which God has given him, he gains knowledge; if he uses them not, he remains ignorant. Mr. R. A. Proctor, who has been called "the greatest astronomer of the age," says: "The Earth on which we live and move seems to be flat." Now, he does not mean that it seems to be flat to the man who shuts his eyes in the face of nature, or, who is not in the full possession of his senses: no, but to the average, common sense, wide-awake, thinking man. He continues: "that is, though there are hills and valleys on its surface, yet it seems to extend on all sides in one and the same general level." Again, he says: "There seems nothing to prevent us from travelling as far as we please in any direction towards the circle all round us, called the *horizon*, where the sky seems to meet the level of the Earth." "The level of the Earth!" Mr. Proctor knows right well what he is talking about, for the book from which we take his words, "Lessons in Elementary Astronomy," was written, he tells us, " to guard the beginner against the captious objections which have from time to time been urged against accepted astronomical theories." The things which are to be defended, then, are these "accepted astronomical theories!" It is not truth that is to be defended against the assaults of error—Oh, no: simply "theories," right or wrong, because they have been "accepted!" Accepted! Why, they have been accepted because it was not thought to be worth while to look at them. Sir John Herschel says: "We shall take for granted, from the outset, the Copernican system of the world." He did not care whether it was the right system or a wrong one, or he would not have done that: he would have looked into it. But, forsooth, the theories are accepted, and, of course, the men who have accepted them are the men who will naturally defend them if they can. So, Richard A. Proctor tries his hand; and we shall see how it fails him. His book was published without any date to it at all. But there is internal evidence which will fix that matter closely enough. We read of the carrying out of the experiments of the celebrated scientist, Alfred R. Wallace, to prove the "convexity" of the surface of standing water, which experiments were conducted in March, 1870, for the purpose of winning Five Hundred Pounds from John Hampden, Esq., of Swindon, England, who had wagered that sum upon the conviction that the said surface is always a level one. Mr. Proctor says: "The experiment was lately tried in a very amusing way." In or about the year 1870, then, Mr. Proctor wrote his book; and, instead of being ignorant of the details of the experiment, he knew all about them. And whether the "amusing" part of the business was the fact that Mr. Wallace wrongfully claimed the five-hundred pounds and got it, or that Mr. Hampden was the victim of the false claim, it is hard to say. The "way" in which the experiment was carried out is, to all intents and purposes, just the way in which Mr. Proctor states that it "can be tried." He says, however, that the distance involved in the experiment "should be three or four miles." Now, Mr. Wallace took up six miles in his experiment, and was unable to prove that there is any "curvature," though he claimed the money and got it; surely it would be "amusing" for anyone to expect to be able to show the "curvature of the earth" in three or four miles, as Mr. Proctor suggests! Nay, it is ridiculous. But "the greatest astronomer of the age" says the thing can be done! And he gives a diagram: "Showing how the roundness of the Earth can be proved by means of three boats on a large sheet of water." (Three or four miles.) But, though the accepted astronomical theories be scattered to the winds, we charge Mr. Proctor either that he has never made the experiment with the three boats, or, that, if he has, the experiment did NOT prove what he says it will. Accepted theories, indeed! Are they to be bolstered up with absurdity and falsehood? Why, if it were possible to show the two ends of a four-mile stretch of water to be on a level, with the centre portion of that water bulged up, the surface of the Earth would be a series of four-mile curves!

But Mr. Proctor says: "We can set three boats in a line on the water, as at A, B, and C, (Fig. 7). Then, if equal masts are placed in these boats, and we place a telescope, as shown, so that when we look through it we see the tops of the masts of A and C, we find the top of the mast B is above the line of sight." Now, here is the point: Mr. Proctor either knows or he ought to know that we shall NOT find anything of the sort! If he has ever tried the experiment, he knows that the three masts will range in a straight line, just as common sense tells us they will. If he has not tried the experiment, he should have tried it, or have paid attention to the details of experiments by those who have tried similar ones a score of times and again. Mr. Proctor may take either horn of the dilemma he pleases: he is just as wrong as a man can be, either way. He mentions no names, but he says: "A person had written a book, in which he said that he had tried such an experiment as the above, and had found that the surface of the water was *not* curved." That person was "PARALLAX," the founder of the Zetetic Philosophy. He continues: "Another person seems to have believed the first, and became so certain that the Earth is flat as to wager a large sum of money that if three boats were placed as in Fig. 7, the middle one would not be above the line joining the two others." That person was John Hampden. And, says Mr. Proctor, "Unfortunately for him, some one who had more sense agreed to take his wager, and, of course, won his money." Now, the "some one who had more sense" was Mr. Wallace. And, says Proctor, in continuation: "He [Hampden?] was rather angry; and it is a strange thing that he was not angry with himself for being so foolish, or with the person who said he had tried the experiment (and so led him astray), but with the person who had won his money!" Here, then, we see that Mr. Proctor knows better than to say that the experiments conducted by "PARALLAX" were things of the imagination only, or that a wrong account had been given of them; and it would be well if he knew better than to try to make his readers believe that either one or the other of these things is the fact. But, there is the Old Bedford Canal now; and there are ten thousand places where the experiment may be tried! Who, then, are the "foolish" people: those who "believe" the record of experiments made by searchers after Truth, or those who shut their eyes to them, throw a doubt upon the record, charge the conductors of the experiments with dishonesty, never conduct similar experiments themselves, and declare the result of such experiments to be so and so, when the declaration can be proved to be false by any man, with a telescope, in twenty-four hours?

Mr. Proctor:—The sphericity of the Earth CANNOT be proved in the way in which you tell us it "can" be! We tell you to take back your words and remodel them on the basis of Truth. Such careless misrepresentations of facts are a disgrace to science—they are the disgrace of theoretical science to-day! Mr. Blackie, in his work on "Self Culture," says: "All flimsy, shallow, and superficial work, in fact, is a lie, of which a man ought to be ashamed."

That the Earth is an extended plane, stretched out in all directions away from the central North, over which hangs, for ever, the North Star, is a fact which all the falsehoods that can be brought to bear upon it with their dead weight will never overthrow: it is God's Truth the face of which, however, man has the power to smirch all over with his unclean hands. Mr. Proctor says: "We learn from astronomy that all these ideas, natural though they seem, are mistaken." Man's natural ideas and conclusions and experimental results are, then, to be overthrown by—what! By "astronomy?" By a thing without a soul—a mere theoretical abstraction, the outcome of the dreamer? Never! The greatest astronomer of the age is not the man, even, who can so much as attempt to manage the business. "We find," says Mr. Proctor, "that the Earth is not flat, but a globe; not fixed, but in very rapid motion; not much larger than the moon, and far smaller than the Sun and the greater number of the stars."

First, then, Mr. Proctor, tell us HOW you find that the Earth is not flat, but a globe! It does not matter that "we find" it so put down in that conglomeration of suppositions which you seek to defend: the question is, What is the evidence of it?—where can it be obtained? "The Earth on which we live and move seems to be flat," you tell us: where, then, is the mistake? If the Earth seem to be what it is not, how are we to trust our senses? And if it is said that we cannot do so, are we to believe it, and consent to be put down lower than the brutes? No, sir: we challenge you, as we have done many times before, to produce the slightest evidence of the Earth's rotundity, from the world of facts around you. You have given to us the statement we have quoted, and we have the right to demand a proof; and if this is not forthcoming, we have before us the duty of denouncing the absurd dogma as worse than an absurdity—as a FRAUD—and as a fraud that flies in the face of divine revelation! Well, then, Mr. Proctor, in demanding a proof of the Earth's rotundity (or the frank admission of your errors), we are tempted to taunt you as we tell you that it is utterly out of your power to produce one; and we tell you that you do not dare even to lift up your finger to point us to the so-called proofs in the school-books of the day, for you know the measure of absurdity of which they are composed, and how disgraceful it is to allow them to remain as false guides of the youthful mind!

Mr. Proctor: we charge you that, whilst you teach the theory of the Earth's rotundity and mobility, you KNOW that it is a plane; and here is the ground of the charge. In page 7, in your book, you give a diagram of the "surface on which we live," and the "supposed globe" —the supposed "hollow globe"—of the heavens, arched over the said surface. Now, Mr. Proctor, you picture the surface on which we live in exact accordance with your verbal description. And what is that description? We shall scarcely be believed when we say that we give it just as it stands: "The level of the surface on which we live." And, that there may be no mistake about the meaning of the word "level," we remind you that your diagram proves that the level that you mean is the level of the mechanic, a plane surface, and not the "level" of the astronomer, which is a convex surface! In short, your description of the Earth is exactly what you say it "seems to be," and, yet, what you say it is not: the very aim of your book being to say so! And we call this the prostitution of the printing press. And it is all the evidence that is necessary to bring the charge home to you, since the words and the diagram are in page 7 of your own book. You know, then, that Earth is a Plane—and so do we.

Now for the evidence of this grand fact, that other people may know it as well as you: remembering, from first to last, that you have not dared to bring forward a single item from the mass of evidence which is to be found in the "Zetetic Philosophy," by "Parallax," a work the influence of which it was the avowed object of your own book to crush!—except that of the three boats, an experiment which you have never tried, and the result of which has never been known, by anyone who has tried it, to be as you say it is!

1. The aeronaut can see for himself that Earth is a Plane. The appearance presented to him, even at the highest elevation he has ever attained, is that of a concave surface—this being exactly what is to be expected of a surface that is truly level, since it is the nature of level surfaces to appear to rise to a level with the eye of the observer. This is ocular demonstration and proof that Earth is not a globe.

2. Whenever experiments have been tried on the surface of standing water, this surface has always been found to be level. If the Earth were a globe, the surface of all standing water would be convex. This is an experimental proof that Earth is not a globe.

3. Surveyors' operations in the construction of railroads, tunnels, or canals are conducted without the slightest "allowance" being made for "curvature," although it is taught that this so-called allowance is absolutely necessary! This is a cutting proof that Earth is not a globe.

4. There are rivers that flow for hundreds of miles towards the level of the sea without falling more than a few feet—notably, the Nile, which, in a thousand miles, falls but a foot. A level expanse

of this extent is quite incompatible with the idea of the Earth's "convexity." It is, therefore, a reasonable proof that Earth is not a globe.

5. The lights which are exhibited in lighthouses are seen by navigators at distances at which, according to the scale of the supposed "curvature" given by astronomers, they ought to be many hundreds of feet, in some cases, down below the line of sight! For instance: the light at Cape Hatteras is seen at such a distance (40 miles) that, according to theory, it ought to be nine-hundred feet higher above the level of the sea than it absolutely is, in order to be visible! This is a conclusive proof that there is no "curvature," on the surface of the sea—" the level of the sea,"—ridiculous though it is to be under the necessity of proving it at all: but it is, nevertheless, a conclusive proof that the Earth is not a globe.

6. If we stand on the sands of the sea-shore and watch a ship approach us, we shall find that she will apparently "rise"—to the extent of her own height, nothing more. If we stand upon an eminence, the same law operates still; and it is but the law of perspective, which causes objects, as they approach us, to appear to increase in size until we see them, close to us, the size they are in fact. That there is no other "rise" than the one spoken of is plain from the fact that, no matter how high we ascend above the level of the sea, the horizon rises on and still on as we rise, so that it is always on a level with the eye, though it be two-hundred miles away, as seen by Mr. J. Glaisher, of England, from Mr. Coxwell's balloon. So that a ship five miles away may be imagined to be "coming up" the imaginary downward curve of the Earth's surface, but if we merely ascend a hill such as Federal Hill, Baltimore, we may see twenty-five miles away, on a level with the eye—that is, twenty miles level distance beyond the ship that we vainly imagined to be "rounding the curve," and "coming up!" This is a plain proof that the Earth is not a globe.

7. If we take a trip down the Chesapeake Bay, in the day-time, we may see for ourselves the utter fallacy of the idea that when a vessel appears "hull down," as it is called, it is because the hull is "behind the water:" for, vessels have been seen, and may often be seen again, presenting the appearance spoken of, and away—far away—beyond those vessels, and, at the same moment, the level shore line, with its accompanying complement of tall trees, towering up, in perspective, over the heads of the "hull-down" ships! Since, then, the idea will not stand its ground when the facts rise up against it, and it is a piece of the popular theory, the theory is a contemptible piece of business, and we may easily wring from it a proof that Earth is not a globe.

8. If the Earth were a globe, a small model globe would be the very best—because the truest—thing for the navigator to take to sea with him. But such a thing as that is not known: with such a toy as a guide, the mariner would wreck his ship, of a certainty! This is a proof that Earth is not a globe.

9. As mariners take to sea with them charts constructed as though the sea were a level surface, however these charts may err as to the true form of this level surface taken as a whole, it is clear, as they find them answer their purpose tolerably well—and only tolerably well, for many ships are wrecked owing to the error of which we

speak—that the surface of the sea is as it is taken to be, whether the captain of the ship "supposes" the Earth to be a globe or anything else. Thus, then, we draw, from the common system of "plane sailing," a practical proof that Earth is not a globe.

10. That the mariners' compass points north and south at the same time is a fact as indisputable as that two and two makes four; but that this would be impossible if the thing were placed on a globe with "north" and "south" at the centre of opposite hemispheres is a fact that does not figure in the school-books, though very easily seen: and it requires no lengthy train of reasoning to bring out of it a pointed proof that the Earth is not a globe.

11. As the mariners' compass points north and south at one time, and as the North, to which it is attracted, is that part of the Earth situate where the North Star is in the zenith, it follows that there is no south "point" or "pole" but that, while the centre is North, a vast circumference must be South in its whole extent. This is a proof that the Earth is not a globe.

12. As we have seen that there is, really, no south point (or pole) but an infinity of points forming, together, a vast circumference—the boundary of the known world, with its battlements of icebergs which bid defiance to man's onward course in a southerly direction—so there can be no east or west "points,'" just as there is no "yesterday," and no "to-morrow." In fact, as there is one point that is fixed (the North), it is impossible for any other point to be fixed likewise. East and west are, therefore, merely directions at right angles with a north and south line: and as the south point of the compass shifts round to all parts of the circular boundary, (as it may be carried round the central North), so the directions east and west, crossing this line, continued, form a circle, at any latitude. A westerly circumnavigation, therefore, is a going round with the North Star continually on the right hand, and an easterly circumnavigation is performed only when the reverse condition of things is maintained, the North Star being on the left hand as the journey is made. These facts, taken together, form a beautiful proof that the Earth is not a globe.

13. As the mariners' compass points north and south at one and the same time, and a meridian is a north and south line, it follows that meridians can be no other than straight lines. But, since all meridians on a globe are semicircles, it is an incontrovertible proof that the Earth is not a globe.

14. "Parallels of latitude" only—of all imaginary lines on the surface of the Earth—are circles, which increase, progressively, from the northern centre to the southern circumference. The mariner's course in the direction of any one of these concentric circles is his longitude, the degrees of which INCREASE to such an extent beyond the equator (going southwards) that hundreds of vessels have been wrecked because of the false idea created by the untruthfulness of the charts and the globular theory together, causing the sailor to be continually getting out of his reckoning. With a map of the Earth in its true form all difficulty is done away with, and ships may be conducted anywhere with perfect safety. This, then, is a very important practical proof that the Earth is not a globe.

15. The idea that, instead of sailing horizontally round the Earth, ships are taken down one side of a globe, then underneath, and are brought up on the other side to get home again, is, except as a mere dream, impossible and absurd! And, since there are neither impossibilities nor absurdities in the simple matter of circumnavigation, it stands, without argument, a proof that the Earth is not a globe.

16. If the Earth were a globe, the distance round its surface at, say, 45 "degrees" south latitude, could not possibly be any greater than it is at the same latitude north; but, since it is found by navigators to be twice the distance—to say the least of it—or, double the distance it ought to be according to the globular theory, it is a proof that the Earth is not a globe.

17. Human beings require a surface on which to live that, in its general character, shall be LEVEL; and since the Omniscient Creator must have been perfectly acquainted with the requirements of His creatures, it follows that, being an All-wise Creator, He has met them thoroughly. This is a theological proof that the Earth is not a globe.

18. The best possessions of man are his senses; and, when he uses them all, he will not be deceived in his survey of nature. It is only when some one faculty or other is neglected or abused that he is deluded. Every man in full command of his senses knows that a level surface is a flat or horizontal one; but astronomers tell us that the true level is the curved surface of a globe! They know that man requires a level surface on which to live, so they give him one in name which is not one in fact! Since this is the best that astronomers, with their theoretical science, can do for their fellow creatures—deceive them—it is clear that things are not as they say they are; and, in short, it is a proof that Earth is not a globe.

19. Every man in his senses goes the most reasonable way to work to do a thing. Now, astronomers (one after another—following a leader), while they are telling us that Earth is a globe, are cutting off the upper half of this supposititious globe in their books, and, in this way, forming the level surface on which they describe man as living and moving! Now, if the Earth were really a globe, this would be just the most unreasonable and suicidal mode of endeavoring to show it. So that, unless theoretical astronomers are all out of their senses together, it is, clearly, a proof that the Earth is not a globe.

20. The common sense of man tells him—if nothing else told him—that there is an "up" and a "down" in nature, even as regards the heavens and the earth; but the theory of modern astronomers necessitates the conclusion that there is not: therefore, the theory of the astronomers is opposed to common sense—yes, and to inspiration—and this is a common sense proof that the Earth is not a globe.

21. Man's experience tells him that he is not constructed like the flies that can live and move upon the ceiling of a room with as much safety as on the floor: and since the modern theory of a planetary earth necessitates a crowd of theories to keep company with it, and one of them is that men are really bound to the earth by a force which fastens them to it "like needles round a spherical loadstone," a

theory perfectly outrageous and opposed to all human experience, it follows that, unless we can trample upon common sense and ignore the teachings of experience, we have an evident proof that the Earth is not a globe.

22. God's Truth never—no, never—requires a falsehood to help it along. Mr. Proctor, in his "Lessons," says: Men "have been able to go round and round the Earth in several directions." Now, in this case, the word "several" will imply more than two, unquestionably: whereas, it is utterly impossible to circumnavigate the Earth in any other than an easterly or a westerly direction; and the fact is perfectly consistent and clear in its relation to Earth as a Plane. Now, since astronomers would not be so foolish as to damage a good cause by misrepresentation, it is presumptive evidence that their cause is a bad one, and—a proof that Earth is not a globe.

23. If astronomical works be searched through and through, there will not be found a single instance of a bold, unhesitating, or manly statement respecting a proof of the Earth's "rotundity." Proctor speaks of "proofs which serve to show .. that the Earth is not flat," and says that man "finds reason to think that the Earth is not flat," and speaks of certain matters being "explained by supposing" that the Earth is a globe; and says that people have "assured themselves that it is a globe;" but he says, also, that there is a "most complete proof that the Earth is a globe:" just as though anything in the world could possibly be wanted but a proof—a proof that proves and settles the whole question. This, however, all the money in the United States Treasury would not buy; and, unless the astronomers are all so rich that they don't want the cash, it is a sterling proof that the Earth is not a globe.

24. When a man speaks of a "most complete" thing amongst several other things which claim to be what that thing is, it is evident that they must fall short of something which the "most complete" thing possesses. And when it is known that the "most complete" thing is an entire failure, it is plain that the others, all and sundry, are worthless. Proctor's "most complete proof that the Earth is a globe" lies in what he calls "the fact" that distances from place to place agree with calculation. But, since the distance round the Earth at 45 "degrees." south of the equator is twice the distance it would be on a globe, it follows that what the greatest astronomer of the age calls "a fact" is NOT a fact; that his "most complete proof" is a most complete failure; and that he might as well have told us, at once, that he has NO PROQF to give us at all. Now, since, if the Earth be a globe, there would, necessarily, be piles of proofs of it all round us, it follows that when astronomers, with all their ingenuity, are utterly unable to point one out—to say nothing about picking one up—that they give us a proof that Earth is not a globe.

25. The surveyor's plans in relation to the laying of the first Atlantic Telegraph cable, show that in 1665 miles—from Valentia, Ireland, to St. John's, Newfoundland—the surface of the Atlantic Ocean is a LEVEL surface—not the astronomers' "level," either! The authoritative drawings, published at the time, are a standing evidence of the fact, and form a practical proof that Earth is not a globe.

EARTH IS NOT A GLOBE.   11

26. If the Earth were a globe, it would, if we take Valentia to be the place of departure, curvate downwards, in the 1665 miles across the Atlantic to Newfoundland, according to the astronomers' own tables, more than three-hundred miles; but, as the surface of the Atlantic does not do so—the fact of its levelness having been clearly demonstrated by Telegraph Cable surveyors,—it follows that we have a grand proof that Earth is not a globe.

27. Astronomers, in their consideration of the supposed "curvature" of the Earth, have carefully avoided the taking of that view of the question which—if anything were needed to do so—would show its utter absurdity. It is this:—If, instead of taking our ideal point of departure to be at Valentia, we consider ourselves at St. John's, the 1665 miles of water between us and Valentia would just as well "curvate" downwards as it did in the other case! Now, since the direction in which the Earth is said to "curvate" is interchangeable—depending, indeed, upon the position occupied by a man upon its surface—the thing is utterly absurd; and it follows that the theory is an outrage, and that the Earth does not "curvate" at all:—an evident proof that the Earth is not a globe.

28. Astronomers are in the habit of considering two points on the Earth's surface, without, it seems, any limit as to the distance that lies between them, as being on a level, and the intervening section, even though it be an ocean, as a vast "hill"—of water! The Atlantic ocean, in taking this view of the matter, would form a "hill of water" more than a hundred miles high! The idea is simply monstrous, and could only be entertained by scientists whose whole business is made up of materials of the same description: and it certainly requires no argument to deduce, from such "science" as this, a satisfactory proof that the Earth is not a globe.

29. If the Earth were a globe, it would, unquestionably, have the same general characteristics—no matter its size—as a small globe that may be stood upon the table. As the small globe has top, bottom, and sides, so must also the large one—no matter how large it be. But, as the Earth, which is "supposed" to be a large globe, has no sides or bottom as the small globe has, the conclusion is irresistible that it is a proof that Earth is not a globe.

30. If the Earth were a globe, an observer who should ascend above its surface would have to look downwards at the horizon (if it be possible to conceive of a horizon at all under such circumstances) even as astronomical diagrams indicate—at angles varying from ten to nearly fifty degrees below the "horizontal" line of sight! (It is just as absurd as it would be to be taught that when we look at a man full in the face we are looking down at his feet!) But, as no observer in the clouds, or upon any eminence on the earth, has ever had to do so, it follows that the diagrams spoken of are imaginary and false; that the theory which requires such things to prop it up is equally airy and untrue; and that we have a substantial proof that Earth is not a globe.

31. If the Earth were a globe, it would certainly have to be as large as it is said to be—twenty-five thousand miles in circumference. Now, the thing which is called a "proof" of the Earth's roundness, and which is presented to children at school, is, that if we stand on

the sea-shore we may see the ships, as they approach us, absolutely "coming up," and that, as we are able to see the highest parts of these ships first, it is because the lower parts are "behind the earth's curve." Now, since, if this were the case—that is, if the lower parts of these ships were behind a "hill of water" at all—the size of the Earth, indicated by such a curve as this, would be so small that it would only be big enough to hold the people of a parish, if they could get all round it, instead of the nations of the world, it follows that the idea is preposterous; that the appearance is due to another and to some reasonable cause; and that, instead of being a proof of the globular form of the Earth, it is a proof that Earth is not a globe.

32. It is often said that, if the Earth were flat, we could see all over it! This is the result of ignorance. If we stand on the level surface of a plain or a prairie, and take notice, we shall find that the horizon is formed at about three miles all around us: that is, the ground appears to rise up until, at that distance, it seems on a level with the eye-line or line of sight. Consequently, objects no higher than we stand—say, six feet—and which are at that distance (three miles), have reached the "vanishing point," and are beyond the sphere of our unaided vision. This is the reason why the hull of a ship disappears (in going away from us) before the sails; and, instead of there being about it the faintest shadow of evidence of the Earth's rotundity, it is a clear proof that Earth is not a globe.

33. If the Earth were a globe, people—except those on the top—would, certainly, have to be "fastened" to its surface by some means or other, whether by the "attraction" of astronomers or by some other undiscovered and undiscoverable process! But, as we know that we simply walk on its surface without any other aid than that which is necessary for locomotion on a plane, it follows that we have, herein, a conclusive proof that Earth is not a globe.

34. If the Earth were a globe, there certainly would be—if we could imagine the thing to be peopled all round—"antipodes:" "people who," says the dictionary, "living exactly on the opposite side of the globe to ourselves, have their feet opposite to ours:" —people who are hanging heads downwards whilst we are standing heads up! But, since the theory allows us to travel to those parts of the Earth where the people are said to be heads downwards, and still to fancy ourselves to be heads upwards and our friends whom we have left behind us to be heads downwards, it follows that the whole thing is a myth—a dream—a delusion—and a snare; and, instead of there being any evidence at all in this direction to substantiate the popular theory, it is a plain proof that the Earth is not a globe.

35. If we examine a true picture of the distant horizon, or the thing itself, we shall find that it coincides exactly with a perfectly straight and level line. Now, since there could be nothing of the kind on a globe, and we find it to be the case all over the Earth, it is a proof that the Earth is not a globe.

36. If we take a journey down the Chesapeake Bay, by night, we shall see the "light" exhibited at Sharpe's Island for an hour before the steamer gets to it. We may take up a position on the deck

so that the rail of the vessel's side will be in a line with the "light" and in the line of sight; and we shall find that in the whole journey the light will not vary in the slightest degree in its apparent elevation. But, say that a distance of thirteen miles has been traversed, the astronomers' theory of "curvature" demands a difference (one way or the other!) in the apparent elevation of the light, of 112 feet 8 inches! Since, however, there is not a difference of 112 hair's breadths, we have a plain proof that the water of the Chesapeake Bay is not curved, which is a proof that the Earth is not a globe.

37. If the Earth were a globe, there would, very likely, be (for nobody knows) six months day and six months night at the arctic and antarctic regions, as astronomers dare to assert there is:—for their theory demands it! But, as this fact—the six months day and six months night—is nowhere found but in the arctic regions, it agrees perfectly with everything else that we know about the Earth as a plane, and, whilst it overthrows the "accepted theory," it furnishes a striking proof that Earth is not a globe.

38. When the Sun crosses the equator, in March, and begins to circle round the heavens in north latitude, the inhabitants of high northern latitudes see him skimming round their horizon and forming the break of their long day, in a horizontal course, not disappearing again for six months, as he rises higher and higher in the heavens whilst he makes his twenty-four hour circle until June, when he begins to descend and goes on until he disappears beyond the horizon in September. Thus, in the northern regions, they have that which the traveller calls the "midnight Sun," as he sees that luminary at a time when, in his more southern latitude, it is always midnight. If, then, for one-half the year, we may see for ourselves the Sun making horizontal circles round the heavens, it is presumptive evidence that, for the other half-year, he is doing the same, although beyond the boundary of our vision. This, being a proof that Earth is a plane, is, therefore, a proof that the Earth is not a globe.

39. We have abundance of evidence that the Sun moves daily round and over the Earth in circles concentric with the northern region over which hangs the North Star; but, since the theory of the Earth being a globe is necessarily connected with the theory of its motion round the Sun in a yearly orbit, it falls to the ground when we bring forward the evidence of which we speak, and, in so doing, forms a proof that the Earth is not a globe.

40. The Suez Canal, which joins the Red Sea with the Mediterranean, is about one hundred miles long; it forms a straight and level surface of water from one end to the other; and no "allowance" for any supposed "curvature" was made in its construction. It is a clear proof that the Earth is not a globe.

41. When astronomers assert that it is "necessary" to make "allowance for curvature" in canal construction, it is, of course, in order that, in their idea, a level cutting may be had for the water. How flagrantly, then, do they contradict themselves when they say that the curved surface of the Earth is a "true level!" What more can they want for a canal than a true level? Since they contradict themselves in such an elementary point as this, it is an evidence that the whole

thing is a delusion, and we have a proof that the Earth is not a globe.

42. It is certain that the theory of the Earth's rotundity and that of its mobility must stand or fall together. A proof, then, of its immobility is virtually a proof of its non-rotundity. Now, that the Earth does not move, either on an axis, or in an orbit round the Sun or anything else, is easily proven. If the Earth went through space at the rate of eleven-hundred miles in a minute of time, as astronomers teach us, in a particular direction, there would unquestionably be a difference in the result of firing off a projectile in that direction and in a direction the opposite of that one. But as, in fact, there is not the slightest difference in any such case, it is clear that any alleged motion of the Earth is disproved, and that, therefore, we have a proof that the Earth is not a globe.

43. The circumstances which attend bodies which are caused merely to fall from a great height prove nothing as to the motion or stability of the Earth, since the object, if it be on a thing that is in motion, will participate in that motion; but, if an object be thrown upwards from a body at rest, and, again, from a body in motion, the circumstances attending its descent will be very different. In the former case, it will fall, if thrown vertically upwards, at the place from whence it was projected; in the latter case, it will fall behind —the moving body from which it is thrown will leave it in the rear. Now, fix a gun, muzzle upwards, accurately, in the ground; fire off a projectile; and it will fall by the gun. If the Earth travelled eleven-hundred miles a minute, the projectile would fall behind the gun, in the opposite direction to that of the supposed motion. Since, then, this is NOT the case, in fact, the Earth's fancied motion is negatived, and we have a proof that the Earth is not a globe.

44. It is in evidence that, if a projectile be fired from a rapidly moving body in an opposite direction to that in which the body is going, it will fall short of the distance at which it would reach the ground if fired in the direction of motion. Now, since the Earth is said to move at the rate of nineteen miles in a second of time, "from west to east," it would make all the difference imaginable if the gun were fired in an opposite direction. But, as, in practice, there is not the slightest difference, whichever way the thing may be done, we have a forcible overthrow of all fancies relative to the motion of the Earth, and a striking proof that the Earth is not a globe.

45. The Astronomer Royal, of England, George B, Airy, in his celebrated work on Astronomy, the "Ipswich Lectures," says: "Jupiter is a large planet that turns on his axis, and why do not we turn?" Of course, the common sense reply is: Because the Earth is not a planet! When, therefore, an astronomer royal puts words into our mouth wherewith we may overthrow the supposed planetary nature of the Earth, we have not far to go to pick up a proof that Earth is not a globe.

46. It has been shown that an easterly or a westerly motion is necessarily a circular course round the central North. The only north point or centre of motion of the heavenly bodies known to man is that formed by the North Star, which is over the central portion of the outstretched Earth. When, therefore, astronomers tell us of

a planet taking a westerly course round the Sun, the thing is as meaningless to them as it is to us, unless they make the Sun the northern centre of the motion, which they cannot do! Since, then, the motion which they tell us the planets have is, on the face of it, absurd; and since, as a matter of fact, the Earth can have no absurd motion at all, it is clear that it cannot be what astronomers say it is—a planet; and, if not a planet, it is a proof that Earth is not a globe.

47. In consequence of the fact being so plainly seen, by everyone who visits the sea-shore, that the line of the horizon is a perfectly straight line, it becomes impossible for astronomers, when they attempt to convey, pictorially, an idea of the Earth's "convexity," to do so with even a shadow of consistency: for they dare not represent this horizon as a curved line, so well known is it that it is a straight one! The greatest astronomer of the age, in page 15 of his "Lessons," gives an illustration of a ship sailing away, "as though she were rounding the top of a great hill of water;" and there—of a truth—is the straight and level line of the horizon clear along the top of the "hill" from one side of the picture to the other! Now, if this picture were true in all its parts—and it is outrageously false in several—it would show that Earth is a cylinder; for the "hill" shown is simply up one side of the level, horizontal line, and, we are led to suppose, down the other! Since, then, we have such high authority as Professor Richard A. Proctor that the Earth is a cylinder, it is, certainly, a proof that the Earth is not a globe.

48. In Mr. Proctor's "Lessons in Astronomy," page 15, a ship is represented as sailing away from the observer, and it is given in five positions or distances away on its journey. Now, in its first position, its mast appears above the horizon, and, consequently, higher than the observer's line of vision. But, in its second and third positions, representing the ship as further and further away, it is drawn higher and still higher up above the line of the horizon! Now, it is utterly impossible for a ship to sail away from an observer, under the conditions indicated, and to appear as given in the picture. Consequently, the picture is a misrepresentation, a fraud, and a disgrace. A ship starting to sail away from an observer with her masts above his line of sight would appear, indisputably, to go down and still lower down towards the horizon line, and could not possibly appear—to anyone with his vision undistorted—as going in any other direction, curved or straight. Since, then, the design of the astronomer-artist is to show the Earth to be a globe, and the points in the picture, which would only prove the Earth to be cylindrical if true, are NOT true, it follows that the astronomer-artist fails to prove, pictorially, either that the Earth is a globe or a cylinder, and that we have, therefore, a reasonable proof that the Earth is not a globe.

49. It is a well-known fact that clouds are continually seen moving in all manner of directions—yes, and frequently, in different directions at the same time—from west to east being as frequent a direction as any other. Now, if the Earth were a globe, revolving through space from west to east at the rate of nineteen miles in a second, the clouds appearing to us to move towards the east would have to move quicker than nineteen miles in a second to be thus seen; whilst those which

appear to be moving in the opposite direction would have no necessity to be moving at all, since the motion of the Earth would be more than sufficient to cause the appearance. But it only takes a little common sense to show us that it is the clouds that move just as they appear to do, and that, therefore, the Earth is motionless. We have, then, a proof that the Earth is not a globe.

50. We read in the inspired book, or collection of books, called THE BIBLE, nothing at all about the Earth being a globe or a planet, from beginning to end, but hundreds of allusions there are in its pages which could not be made if the Earth were a globe, and which are, therefore, said by the astronomer to be absurd and contrary to what he knows to be true! This is the groundwork of modern infidelity. But, since every one of many, many allusions to the Earth and the heavenly bodies in the Scriptures can be demonstrated to be absolutely true to nature, and we read of the Earth being "stretched out" "above the waters," as "standing in the water and out of the water," of its being "established that it cannot be moved," we have a store from which to take all the proofs we need, but we will just put down one proof—the Scriptural proof—that Earth is not a globe.

51. A "Standing Order" exists in the English Houses of Parliament that, in the cutting of canals, &c., the datum line employed shall be a "horizontal line, which shall be the same throughout the whole length of the work." Now, if the Earth were a globe, this "Order" could not be carried out: but, it is carried out: therefore, it is a proof that the Earth is not a globe.

52. It is a well-known and indisputable fact that there is a far greater accumulation of ice south of the equator than is to be found at an equal latitude north: and it is said that at Kerguelen, 50 degrees south, 18 kinds of plants exist, whilst, in Iceland, 15 degrees nearer the northern centre, there are 870 species; and, indeed, all the facts in the case show that the Sun's power is less intense at places in the southern region than it is in corresponding latitudes north. Now, on the Newtonian hypothesis, all this is inexplicable, whilst it is strictly in accordance with the facts brought to light by the carrying out of the principles involved in the Zetetic Philosophy of "Parallax." This is a proof that the Earth is not a globe.

53. Every year the Sun is as long south of the equator as he is north; and if the Earth were not "stretched out" as it is, in fact, but turned under, as the Newtonian theory suggests, it would certainly get as intensive a share of the Sun's rays south as north; but the Southern region being, in consequence of the fact stated, far more extensive than the region North, the Sun, having to complete his journey round every twenty-four hours, travels quicker as he goes further south, from September to December, and his influence has less time in which to accumulate at any given point. Since, then, the facts could not be as they are if the Earth were a globe, it is a proof that the Earth is not a globe.

54. The aeronaut is able to start in his balloon and remain for hours in the air, at an elevation of several miles, and come down again in the same county or parish from which he ascended. Now, unless the Earth drag the balloon along with it in its nineteen-miles-a-

second motion, it must be left far behind, in space: but, since balloons have never been known thus to be left, it is a proof that the Earth does not move, and, therefore, a proof that the Earth is not a globe.

55. The Newtonian theory of astronomy requires that the Moon "borrow" her light from the Sun. Now, since the Sun's rays are hot and the Moon's light sends with it no heat at all, it follows that the Sun and Moon are "two great lights," as we somewhere read; that the Newtonian theory is a mistake; and that, therefore, we have a proof that the Earth is not a globe.

56. The Sun and Moon may often be seen high in the heavens at the same time—the Sun rising in the east and the Moon setting in the west—the Sun's light positively putting the Moon's light out by sheer contrast! If the accepted Newtonian theory were correct, and the Moon had her light from the Sun, she ought to be getting more of it when face to face with that luminary—if it were possible for a sphere to act as a reflector all over its face! But as the Moon's light pales before the rising Sun, it is a proof that the theory fails; and this gives us a proof that the Earth is not a globe.

57. The Newtonian hypothesis involves the necessity of the Sun, in the case of a lunar eclipse, being on the opposite side of a globular earth, to cast its shadow on the Moon: but, since eclipses of the Moon have taken place with both the Sun and the Moon above the horizon, it follows that it cannot be the shadow of the Earth that eclipses the Moon; that the theory is a blunder; and that it is nothing less than a proof that the Earth is not a globe.

58. Astronomers have never agreed amongst themselves about a rotating Moon revolving round a rotating and revolving Earth—this Earth, Moon, planets and their satellites all, at the same time dashing through space, around the rotating and revolving Sun, towards the constellation Hercules, at the rate of four millions of miles a day! And they never will: agreement is impossible! With the Earth a plane and without motion, the whole thing is clear. And if a straw will show which way the wind blows, this may be taken as a pretty strong proof that the Earth is not a globe.

59. Mr. Proctor says: "The Sun is so far off that even moving from one side of the Earth to the other does not cause him to be seen in a different direction—at least the difference is too small to be measured." Now, since we know that north of the equator, say 45 degrees, we see the Sun at mid-day to the south, and that at the same distance south of the equator we see the Sun at mid-day to the north, our very shadows on the ground cry aloud against the delusion of the day and give us a proof that Earth is not a globe.

60. There is no problem more important to the astronomer than that of the Sun's distance from the Earth. Every change in the estimate changes everything. Now, since modern astronomers, in their estimates of this distance, have gone all the way along the line of figures from three millions of miles to a hundred and four millions—to-day, the distance being something over 91,000,000; it matters not how much: for, not many years ago, Mr. Hind gave the distance, "accurately," as 95,370,000!—it follows that they don't know, and that it is foolish for anyone to expect that they ever will know, the Sun's

distance! And since all this speculation and absurdity is caused by the primary assumption that Earth is a wandering, heavenly body, and is all swept away by a knowledge of the fact that Earth is a plane, it is a clear proof that Earth is not a globe.

61. It is plain that a theory of measurements without a measuring-rod is like a ship without a rudder; that a measure that is not fixed, not likely to be fixed, and never has been fixed, forms no measuring-rod at all; and that as modern theoretical astronomy depends upon the Sun's distance from the Earth as its measuring-rod, and the distance is not known, it is a system of measurements without a measuring-rod—a ship without a rudder. Now, since it is not difficult to foresee the dashing of this thing upon the rock on which Zetetic astronomy is founded, it is a proof that Earth is not a globe.

62. It is commonly asserted that "the Earth must be a globe because people have sailed round it." Now, since this implies that we can sail round nothing unless it be a globe, and the fact is well known that we can sail round the Earth as a plane, the assertion is ridiculous, and we have another proof that Earth is not a globe.

63. It is a fact not so well known as it ought to be that when a ship, in sailing away from us, has reached the point at which her hull is lost to our unaided vision, a good telescope will restore to our view this portion of the vessel. Now, since telescopes are not made to enable people to see through a "hill of water," it is clear that the hulls of ships are not behind a hill of water when they can be seen through a telescope though lost to our unaided vision. This is a proof that Earth is not a globe.

64. Mr. Glaisher, in speaking of his balloon ascents, says: "The horizon always appeared on a level with the car." Now, since we may search amongst the laws of optics in vain for any principle that would cause the surface of a globe to turn its face upwards instead of downwards, it is a clear proof that the Earth is not a globe.

65. The Rev. D. Olmsted, in describing a diagram which is supposed to represent the Earth as a globe, with a figure of a man sticking out at each side and one hanging head downwards, says: "We should dwell on this point until it appears to us as truly up,"—in the direction given to these figures as it does with regard to a figure which he has placed on the top! Now, a system of philosophy which requires us to do something which is, really, the going out of our minds, by dwelling on an absurdity until we think it is a fact, cannot be a system based on God's truth, which never requires anything of the kind. Since, then, the popular theoretical astronomy of the day requires this, it is evident that it is the wrong thing, and that this conclusion furnishes us with a proof that the Earth is not a globe.

66. It is often said that the predictions of eclipses prove astronomers to be right in their theories. But it is not seen that this proves too much. It is well known that Ptolemy predicted eclipses for six-hundred years, on the basis of a plane Earth, with as much accuracy as they are predicted by modern observers. If, then, the predictions prove the truth of the particular theories current at the time, they just as well prove one side of the question as the other, and enable us to lay claim to a proof that the Earth is not a globe.

67. Seven-hundred miles is said to be the length of the great Canal, in China. Certain it is that, when this canal was formed, no "allowance" was made for "curvature." Yet the canal is a fact without it. This is a Chinese proof that the Earth is not a globe.

68. Mr. J. N. Lockyer says: "Because the Sun seems to rise in the east and set in the west, the Earth really spins in the opposite direction; that is, from west to east." Now, this is no better than though we were to say—Because a man seems to be coming up the street, the street really goes down to the man! And since true science would contain no such nonsense as this, it follows that the so-called science of theoretical astronomy is not true, and, therefore, we have a proof that the Earth is not a globe.

69. Mr. Lockyer says: "The appearances connected with the rising and setting of the Sun and stars may be due either to our earth being at rest and the Sun and stars travelling round it, or the earth itself turning round, while the Sun and stars are at rest." Now, since true science does not allow of any such beggarly alternatives as these, it is plain that modern theoretical astronomy is not true science, and that its leading dogma is a fallacy. We have, then, a plain proof that the Earth is not a globe.

70. Mr. Lockyer, in describing his picture of the supposed proof of the Earth's rotundity by means of ships rounding a "hill of water," uses these words:—"Diagram showing how, when we suppose the earth is round, we explain how it is that ships at sea appear as they do." This is utterly unworthy of the name of Science! A science that begins by supposing, and ends by explaining the supposition, is, from beginning to end, a mere farce. The men who can do nothing better than amuse themselves in this way must be denounced as dreamers only, and their leading dogma a delusion. This is a proof that Earth is not a globe.

71. The astronomers' theory of a globular Earth necessitates the conclusion that, if we travel south of the equator, to see the North Star is an impossibility. Yet it is well known this star has been seen by navigators when they have been more than 20 degrees south of the equator. This fact, like hundreds of other facts, puts the theory to shame, and gives us a proof that the Earth is not a globe.

72. Astronomers tell us that, in consequence of the Earth's "rotundity," the perpendicular walls of buildings are, nowhere, parallel, and that even the walls of houses on opposite sides of a street are not strictly so! But, since all observation fails to find any evidence of this want of parallelism which theory demands, the idea must be renounced as being absurd and in opposition to all well-known facts. This is a proof that the Earth is not a globe.

73. Astronomers have made experiments with pendulums which have been suspended from the interior of high buildings, and have exulted over the idea of being able to prove the rotation of the Earth on its "axis," by the varying direction taken by the pendulum over a prepared table underneath—asserting that the table moved round under the pendulum, instead of the pendulum shifting and oscillating in different directions over the table! But, since it has been found that, as often as not, the pendulum went round the wrong way for the

"rotation" theory, chagrin has taken the place of exultation, and we have a proof of the failure of astronomers in their efforts to substantiate their theory, and, therefore, a proof that Earth is not a globe.

74. As to the supposed "motion of the whole Solar system in space," the Astronomer Royal of England once said: "The matter is left in a most delightful state of uncertainty, and I shall be very glad if anyone can help us out of it." But, since the whole Newtonian scheme is, to-day, in a most deplorable state of uncertainty—for, whether the Moon goes round the Earth or the Earth round the Moon has, for years, been a matter of "raging" controversy—it follows that, root and branch, the whole thing, is wrong; and, all hot from the raging furnace of philosophical phrensy, we find a glowing proof that Earth is not a globe.

75. Considerably more than a million Earths would be required to make up a body like the Sun—the astronomers tell us: and more than 53,000 suns would be wanted to equal the cubic contents of the star Vega. And Vega is a "small star!" And there are countless millions of these stars! And it takes 30,000,000 years for the light of some of these stars to reach us at 12,000,000 miles in a minute! And, says Mr. Proctor, "I think a moderate estimate of the age of the Earth would be 500,000,000 years! "Its weight," says the same individual, "is 6,000,000,000,000,000,000,000 tons!" Now, since no human being is able to comprehend these things, the giving of them to the world is an insult—an outrage. And though they have all arisen from the one assumption that Earth is a planet, instead of upholding the assumption, they drag it down by the weight of their own absurdity, and leave it lying in the dust—a proof that Earth is not a globe.

76. Mr. J. R. Young, in his work on Navigation, says: "Although the path of the ship is on a spherical surface, yet we may represent the length of the path by a straight line on a plane surface." (And plane sailing is the rule.) Now, since it is altogether impossible to "represent" a curved line by a straight one, and absurd to make the attempt, it follows that a straight line represents a straight line and not a curved one. And, since it is the surface of the waters of the ocean that is being considered by Mr. Young, it follows that this surface is a straight surface, and we are indebted to Mr. Young, a professor of navigation, for a proof that the Earth is not a globe.

77. "Oh, but if the Earth is a plane, we could go to the edge and tumble over!" is a very common assertion. This is a conclusion that is formed too hastily, and facts overthrow it. The Earth certainly is, just what man by his observation finds it to be, and what Mr. Proctor himself says it "seems" to be—flat; and we cannot cross the icy barrier which surrounds it. This is a complete answer to the objection, and, of course, a proof that Earth is not a globe.

78. "Yes, but we can circumnavigate the South easily enough," is often said—by those who don't know. The British Ship Challenger recently completed the circuit of the Southern region—indirectly, to be sure—but she was three years about it, and traversed nearly 69,000 miles—a stretch long enough to have taken her six times round on the globular hypothesis. This is a proof that Earth is not a globe.

79. The remark is common enough that we can see the circle of the Earth if we cross the ocean, and that this proves it to be round. Now, if we tie a donkey to a stake on a level common, and he eats the grass all around him, it is only a circular disc that he has to do with, not a spherical mass. Since, then, circular discs may be seen anywhere—as well from a balloon in the air as from the deck of a ship, or from the standpoint of the donkey, it is a proof that the surface of the Earth is a plane surface, and, therefore, a proof that the Earth is not a globe.

80. It is "supposed," in the regular course of the Newtonian theory, that the Earth is, in June, about 190 millions of miles (190,000,000) away from its position in December. Now, since we can, (in middle north latitudes), see the North Star, on looking out of a window that faces it—and out of the very same corner of the very same pane of glass in the very same window—all the year round, it is proof enough for any man in his senses that we have made no motion at all. It is a proof that the Earth is not a globe.

81. Newtonian philosophers teach us that the Moon goes round the Earth from west to east. But observation—man's most certain mode of gaining knowledge—shows us that the Moon never ceases to move in the opposite direction—from east to west. Since, then, we know that nothing can possibly move in two, opposite directions at the same time, it is a proof that the thing is a big blunder; and, in short, it is a proof that the Earth is not a globe.

82. Astronomers tell us that the Moon goes round the Earth in about 28 days. Well, we may see her making her journey round, every day, if we make use of our eyes—and these are about the best things we have to use. The Moon falls behind in her daily motion as compared with that of the Sun to the extent of one revolution in the time specified; but that is not making a revolution. Failing to go as fast as other bodies go in one direction does not constitute a going round in the opposite one—as the astronomers would have us believe! And, since all this absurdity has been rendered necessary for no other purpose than to help other absurdities along, it is clear that the astronomers are on the wrong track; and it needs no long train of reasoning to show that we have a proof that the Earth is not a globe.

83. It has been shown that meridians are, necessarily, straight lines; and that it is impossible to travel round the Earth in a north or south direction: from which it follows that, in the general acceptation of the word "degree,"—the 360th part of a circle—meridians have no degrees: for no one knows anything of a meridian circle or semicircle, to be thus divided. But astronomers speak of degrees of latitude in the same sense as those of longitude. This, then, is done by assuming that to be true which is not true. Zetetic philosophy does not involve this necessity. This proves that the basis of this philosophy is a sound one, and, in short, is a proof that the Earth is not a globe.

84. If we move away from an elevated object on or over a plain or a prairie, the height of the object will apparently diminish as we do so. Now, that which is sufficient to produce this effect on a small scale is sufficient on a large one; and travelling away from an elevated

object, no matter how high, over a level surface, no matter how far, will cause the appearance in question—the lowering of the object. Our modern theoretical astronomers, however, in the case of the apparent lowering of the North Star as we travel southward, assert that it is evidence that the Earth is globular! But, as it is clear that an appearance which is fully accounted for on the basis of known facts cannot be permitted to figure as evidence in favor of that which is only a supposition, it follows that we rightfully order it to stand down, and make way for a proof that the Earth is not a globe.

85. There are rivers which flow east, west, north, and south—that is, rivers are flowing in all directions over the Earth's surface, and at the same time. Now, if the Earth were a globe, some of these rivers would be flowing up-hill and others down, taking it for a fact that there really is an "up" and a "down" in nature, whatever form she assumes. But, since rivers do not flow up-hill, and the globular theory requires that they should, it is a proof that the Earth is not a globe.

86. If the Earth were a globe, rolling and dashing through "space" at the rate of "a hundred miles in five seconds of time," the waters of seas and oceans could not, by any known law, be kept on its surface—the assertion that they could be retained under these circumstances being an outrage upon human understanding and credulity! But as the Earth—that is, the habitable world of dry land—is found to be "standing out of the water and in the water" of the "mighty deep," whose circumferential boundary is ice, we may throw the statement back into the teeth of those who make it and flaunt before their faces the flag of reason and common sense, inscribed with—a proof that the Earth is not a globe.

87. The theory of a rotating and revolving earth demands a theory to keep the water on its surface; but, as the theory which is given for this purpose is as much opposed to all human experience as the one which it is intended to uphold, it is an illustration of the miserable makeshifts to which astronomers are compelled to resort, and affords a proof that the Earth is not a globe.

88. If we could—after our minds had once been opened to the light of Truth—conceive of a globular body on the surface of which human beings could exist, the power—no matter by what name it be called—that would hold them on would, then, necessarily, have to be so constraining and cogent that they could not live; the waters of the oceans would have to be as a solid mass, for motion would be impossible. But we not only exist, but live and move; and the water of the ocean skips and dances like a thing of life and beauty! This is a proof that the Earth is not a globe.

89. It is well known that the law regulating the apparent decrease in the size of objects as we leave them in the distance (or as they leave us) is very different with luminous bodies from what it is in the case of those which are non-luminous. Sail past the light of a small lamp in a row-boat on a dark night, and it will seem to be no smaller when a mile off than it was when close to it. Proctor says, in speaking of the Sun: "his apparent size does not change,"—far off or near. And then he forgets the fact! Mr. Proctor tells us, subsequently, that, if

the traveller goes so far south that the North Star appears on the horizon, "the Sun should therefore look much larger"—if the Earth were a plane! Therefore, he argues, "the path followed cannot have been the straight course,"—but a curved one. Now, since it is nothing but common scientific trickery to bring forward, as an objection to stand in the way of a plane Earth, the non-appearance of a thing which has never been known to appear at all, it follows that, unless that which appears to be trickery were an accident, it was the only course open to the objector—to trick. (Mr. Proctor, in a letter to the "English Mechanic" for Oct. 20, 1871, boasts of having turned a recent convert to the Zetetic philosophy by telling him that his arguments were all very good, but that "it seems as though [mark the language!] the sun ought to look nine times larger in summer." And Mr. Proctor concludes thus: "He saw, indeed, that, in his faith in 'Parallax,' he had 'written himself down an ass.'") Well, then: trickery or no trickery on the part of the objector, the objection is a counterfeit—a fraud—no valid objection at all; and it follows that the system which does not purge itself of these things is a rotten system, and the system which its advocates, with Mr. Proctor at their head, would crush if they could find a weapon to use—the Zetetic philosophy of "Parallax"—is destined to live! This is a proof that the Earth is not a globe.

90. "Is water level, or is it not?" was a question once asked of an astronomer. "Practically, yes; theoretically, no," was the reply. Now, when theory does not harmonize with practice, the best thing to do is to drop the theory. (It is getting too late, now, to say "So much the worse for the facts!") To drop the theory which supposes a curved surface to standing water is to acknowledge the facts which form the basis of Zetetic philosophy. And since this will have to be done—sooner or later,—it is a proof that the Earth is not a globe.

91. "By actual observation," says Schœdler, in his "Book of Nature," "we know that the other heavenly bodies are spherical, hence we unhesitatingly assert that the earth is so also." This is a fair sample of all astronomical reasoning. When a thing is classed amongst "other" things, the likeness between them must first be proven. It does not take a Schœdler to tell us that "heavenly bodies" are spherical, but "the greatest astronomer of the age" will not, now, dare to tell us that THE EARTH is—and attempt to prove it. Now, since no likeness has ever been proven to exist between the Earth and the heavenly bodies, the classification of the Earth with the heavenly bodies is premature—unscientific—false! This is a proof that Earth is not a globe.

92. "There is no inconsistency in supposing that the earth does move round the sun," says the Astronomer Royal of England. Certainly not, when theoretical astronomy is all supposition together! The inconsistency is in teaching the world that the thing supposed is a fact. Since, then, the "motion" of the Earth is supposition only—since, indeed, it is necessary to suppose it at all—it is plain that it is a fiction and not a fact; and, since "mobility" and "sphericity" stand or fall together, we have before us a proof that Earth is not a globe.

93. We have seen that astronomers—to give us a level surface on

which to live—have cut off one-half of the "globe" in a certain picture in their books. [See page 6.] Now, astronomers having done this, one-half of the substance of their "spherical theory" is given up! Since, then, the theory must stand or fall in its entirety, it has really fallen when the half is gone. Nothing remains, then, but a plane Earth, which is, of course, a proof that the Earth is not a globe.

94. In "Cornell's Geography" there is an "Illustrated proof of the Form of the Earth." A curved line on which is represented a ship in four positions, as she sails away from an observer, is an arc of 72 degrees, or one-fifth of the supposed circumference of the "globe"— about 5,000 miles. Ten such ships as those which are given in the picture would reach the full length of the "arc," making 500 miles as the length of the ship. The man, in the picture, who is watching the ship as she sails away, is about 200 miles high; and the tower, from which he takes an elevated view, at least 500 miles high. These are the proportions, then, of men, towers, and ships which are necessary in order to see a ship, in her different positions, as she "rounds the curve" of the "great hill of water" over which she is supposed to be sailing: for, it must be remembered that this supposed "proof" depends upon lines and angles of vision which, if enlarged, would still retain their characteristics. Now, since ships are not built 500 miles long, with masts in proportion, and men are not quite 200 miles high, it is not what it is said to be—a proof of rotundity—but, either an ignorant farce or a cruel piece of deception. In short, it is a proof that the Earth is not a globe.

95. In "Cornell's Intermediate Geography," (1881) page 12, is an "Illustration of the Natural Divisions of Land and Water." This illustration is so nicely drawn that it affords, at once, a striking proof that Earth is a plane. It is true to nature, and bears the stamp of no astronomer-artist. It is a pictorial proof that Earth is not a globe.

96. If we refer to the diagram in "Cornell's Geography," page 4, and notice the ship in its position the most remote from the observer, we shall find that, though it is about 4,000 miles away, it is the same size as the ship that is nearest to him, distant about 700 miles! This is an illustration of the way in which astronomers ignore the laws of perspective. This course is necessary, or they would be compelled to lay bare the fallacy of their dogmas. In short, there is, in this matter, a proof that the Earth is not a globe.

97. Mr. Hind, the English astronomer, says: "The simplicity with which the seasons are explained by the revolution of the Earth in her orbit and the obliquity of the ecliptic, may certainly be adduced as a strong presumptive proof of the correctness"—of the Newtonian theory; "for on no other rational suppositions with respect to the relations of the Earth and Sun, can these and other as well-known phenomena, be accounted for." But, as true philosophy has no "suppositions" at all—and has nothing to do with "suppositions"—and the phenomena spoken of are thoroughly explained by facts, the "presumptive proof" falls to the ground, covered with the ridicule it so richly deserves; and out of the dust of Mr. Hind's "rational suppositions" we see standing before us a proof that Earth is not a globe.

98. Mr. Hind speaks of the astronomer watching a star as it is

"carried across the telescope by the diurnal revolution of the Earth." Now, this is nothing but downright absurdity. No motion of the Earth could possibly carry a star across a telescope or anything else. If the star is carried across anything at all, it is the star that moves, not the thing across which it is carried! Besides, the idea that the Earth, if it were a globe, could possibly move in an orbit of nearly 600,000,000 of miles with such exactitude that the cross-hairs in a telescope fixed on its surface would appear to glide gently over a star "millions of millions" of miles away is simply monstrous; whereas, with a FIXED telescope, it matters not the distance of the stars, though we suppose them to be as far off as the astronomer supposes them to be; for, as Mr. Proctor himself says, "the further away they are, the less they will seem to shift." Why, in the name of common sense, should observers have to fix their telescopes on solid stone bases so that they should not move a hair's-breadth, if the Earth on which they fix them move at the rate of nineteen miles in a second? Indeed, to believe that Mr. Proctor's mass of "six thousand million million million tons" is "rolling, surging, flying, darting on through space for ever" with a velocity compared with which a shot from a cannon is a "very slow coach," with such unerring accuracy that a telescope fixed on granite pillars in an observatory will not enable a lynx-eyed astronomer to detect a variation in its onward motion of the thousandth part of a hair's-breadth is to conceive a miracle compared with which all the miracles on record put together would sink into utter insignificance. Captain R. J. Morrison, the late compiler of "Zadkeil's Almanac," says: "We declare that this 'motion' is all mere 'bosh'; and that the arguments which uphold it are, when examined with an eye that seeks for TRUTH only, mere nonsense, and childish absurdity." Since, then, these absurd theories are of no use to men in their senses, and since there is no necessity for anything of the kind in Zetetic philosophy, it is a "strong presumptive proof"—as Mr. Hind would say—that the Zetetic philosophy is true, and, therefore, a proof that Earth is not a globe.

99. Mr. Hind speaks of two great mathematicians differing only fifty-five yards in their estimate of the Earth's diameter. Why, Sir John Herschel, in his celebrated work, cuts off 480 miles of the same thing to get "round numbers!" This is like splitting a hair on one side of the head and shaving all the hair off on the other! Oh, "science!" Can there be any truth in a science like this? All the exactitude in astronomy is in Practical astronomy—not Theoretical. Centuries of observation have made practical astronomy a noble art and science, based—as we have a thousand times proved it to be—on a fixed Earth; and we denounce this pretended exactitude on one side and the reckless indifference to figures on the other as the basest trash, and take from it a proof that the "science" which tolerates it is a false —instead of being an "exact"—science, and we have a proof that the Earth is not a globe.

100. The Sun, as he travels round over the surface of the Earth, brings "noon" to all places on the successive meridians which he crosses: his journey being made in a westerly direction, places east of the Sun's position have had their noon, whilst places to the west of the

Sun's position have still to get it. Therefore, if we travel easterly, we arrive at those parts of the Earth where "time" is more advanced, the watch in our pocket has to be "put on," or we may be said to "gain time." If, on the other hand, we travel westerly, we arrive at places where it is still "morning," the watch has to be "put back," and it may be said that we "lose time." But, if we travel easterly so as to cross the 180th meridian, there is a loss, there, of a day, which will neutralize the gain of a whole circumnavigation; and, if we travel westerly, and cross the same meridian, we experience the gain of a day, which will compensate for the loss during a complete circumnavigation in that direction. The fact of losing or gaining time in sailing round the world, then, instead of being evidence of the Earth's "rotundity," as it is imagined to be, is, in its practical exemplification, an everlasting proof that the Earth is not a globe.

"And what then?" What then! No intelligent man will ask the question; and he who may be called an intellectual man will know that the demonstration of the fact that the Earth is not a globe is the grandest snapping of the chains of slavery that ever took place in the world of literature or science. The floodgates of human knowledge are opened afresh and an impetus is given to investigation and discovery where all was stagnation, bewilderment and dreams! Is it nothing to know that infidelity cannot stand against the mighty rush of the living water of Truth that must flow on and on until the world shall look "up" once more "to Him that stretched out the earth above the waters"—"to Him that made great lights:—the Sun to rule by day—the Moon and Stars to rule by night?" Is it nothing to know and to feel that the heavenly bodies were made for man, and that the monstrous dogma of an infinity of worlds is overthrown for ever? The old-time English "Family Herald," for July 25, 1885, says, in its editorial, that "The earth's revolution on its own axis was denied, against Galileo and Copernicus, by the whole weight of the Church of Rome." And, in an article on "The Pride of Ignorance," too!—the editor not knowing that if the Earth had an axis to call its "own"—which the Church well knew it had not, and, therefore, could not admit—it would not "revolve" on it; and that the theoretical motion on an axis is that of rotation, and not revolution! Is it nothing to know that "the whole weight of the Church of Rome" was thrown in the right direction, although it has swayed back again like a gigantic pendulum that will regain its old position before long? Is it nothing to know that the "pride of ignorance" is on the other side? Is it nothing to know that, with all the Bradlaughs and Ingersolls of the world telling us to the contrary—Biblical science is true? Is it nothing to know that we are living on a body at rest, and not upon a heavenly body whirling and dashing through space in every conceivable way and with a velocity utterly inconceivable? Is it nothing to know that we can look stedfastly up to Heaven instead of having no heaven to look up to at all? Is it nothing, indeed, to be in the broad daylight of Truth and to be able to go on towards a possible perfection, instead of being wrapped in the darkness of error on the rough ocean of Life, and finding ourselves stranded at last—God alone knows where?

Baltimore, Maryland, U. S. A., August, 1885.

APPENDIX TO THE SECOND EDITION.

The following letters remain unanswered, at the time of going to press, December 7, 1885:—

"71 Chew Street, Baltimore, Nov. 21, 1885. R. A. Proctor, Esq., St. Joe, Mo. Sir: I have sent you two copies of my 'One Hundred Proofs that the Earth is Not a Globe,' and, as several weeks have since elapsed and I have not heard from you, I write to inform you that if you have any remarks to make concerning that publication, and will let me have them in the course of a week or ten days, I will print them—if you say what you may wish to say in about five or six hundred words—in the second edition of the pamphlet, which will very soon be called for. Allow me to say that, as this work is not only 'dedicated' to you but attacks your teachings, the public will be looking for something from your pen very shortly. I hope they may not be disappointed. Yours in the cause of truth, W. Carpenter."

"71 Chew Street, Baltimore, Nov. 24, 1885. Spencer F. Baird, Esq., Secretary of the Smithsonian Institution, Washington, D. C. Sir: —I had the pleasure, several weeks ago, of sending you my 'One Hundred Proofs that the Earth is Not a Globe.' I hope you received them. A second edition is now called for, and I should esteem it a favor if you would write me a few words concerning them that I may print with this forthcoming edition as an appendix to them. If you think any of the 'Hundred Proofs' are unsound, I will print all you may have to say about them, if not over 400 words, as above stated. I have made Richard A. Proctor, Esq., a similar offer, giving him, of course, a little more space. I feel sure that the very great importance of this matter will prompt you to give it your immediate attention. I have the honor to be, sir, yours sincerely, Wm. Carpenter."

Copies of the first edition of this pamphlet have been sent to the leading newspapers of this country and of England, and to very many of the most renowned scientific men of the two countries—from the Astronomer Royal, of England, to Dr. Gilman, of Johns Hopkins University, Baltimore. Several copies have been sent to graduates of different Universities, on application, in consequence of the subjoined advertisement, which has appeared in several newspapers :—

"WANTED.—A Scholar of ripe attainments to review Carpenter's 'One Hundred Proofs that the Earth is Not a Globe.' Liberal remuneration offered. Apply to Wm. Carpenter, 71 Chew Street, Baltimore. N. B.—No one need apply who has not courage enough to append his name to the Review for publication."

☞ We should be pleased to hear from some of the gentlemen in time for the insertion of their courageous attacks in the Third edition!

OPINIONS OF THE PRESS.

"This can only be described as an extraordinary book. . . His arguments are certainly plausible and ingenious, and even the reader who does not agree with him will find a singular interest and fascination in analyzing the 'one hundred proofs.' . . The proofs are set forth in brief, forcible, compact, very clear paragraphs, the meaning of which can be comprehended at a glance."—Daily News, Sept. 24,

28   ONE HUNDRED PROOFS.

"Throughout the entire work there are discernible traces of a strong and reliant mind, and such reliance as can only have been acquired by unbiassed observation, laborious investigation, and final conviction; and the masterly handling of so profound a theme displays evidence of grave and active researches. There is no groping wildly about in the vagueness of theoretical speculations, no empty hypotheses inflated with baseless assertions and false illustrations, but the practical and perspicuous conclusions of a mind emancipated from the prevailing influences of fashionable credence and popular prejudice, and subordinate only to those principles emanating from reason and common sense."—H. D. T., Woodberry News, Sept. 26, 1885.

"We do not profess to be able to overthrow any of his 'Proofs.' And we must admit, and our readers will be inclined to do the same, that it is certainly a strange thing that Mr. Wm. Carpenter, or anyone else, should be able to bring together 'One Hundred Proofs of anything in the world if that thing is not right, while we keep on asking for one proof, that is really a satisfactory one, on the other side. If these 'Hundred Proofs' are nonsense, we cannot prove them to be so, and some of our scientific men had better try their hands, and we think they will try their heads pretty badly into the bargain."—The Woodberry News, Baltimore, Sept. 19, 1885.

"This is a remarkable pamphlet. The author has the courage of his convictions, and presents them with no little ingenuity, however musty they may appear to nineteenth century readers. He takes for his text a statement of Prof. Proctor's that 'The Earth on which we live and move seems to be flat,' and proceeds with great alacrity to marshal his hundred arguments in proof that it not only seems but is flat, 'an extended plane, stretched out in all directions away from the central North.' He enumerates all the reasons offered by scientists for a belief in the rotundity of the earth and evidently to his own complete satisfaction refutes them. He argues that the heavenly bodies were made solely to light this world, that the belief in an infinity of worlds is a monstrous dogma, contrary to Bible teaching, and the great stronghold of the infidel; and that the Church of Rome was right when it threw the whole weight of its influence against Galileo and Copernicus when they taught the revolution of the earth on its axis."—Michigan Christian Herald, Oct. 15, 1885.

"So many proofs."—Every Saturday, Sept. 26, 1885.

"A highly instructive and very entertaining work. . . The book is well worth reading."—Protector, Baltimore, Oct. 3, 1885.

"The book will be sought after and read with peculiar interest."—Baltimore Labor Free Press, Oct. 17, 1885.

"Some of them [the proofs] are of sufficient force to demand an answer from the advocates of the popular theory."—Baltimore Episcopal Methodist, October 28, 1885.

"Showing considerable smartness both in conception and argument."—Western Christian Advocate, Cincinnati, O., Oct. 21, 1885.

"Forcible and striking in the extreme."—Brooklyn Market Journal.

Baltimore, Maryland, U. S. A., December 7, 1885.

[Appendix to Third Edition.]
COPY OF LETTER FROM RICHARD A. PROCTOR, ESQ.
5 Montague Street, Russell Square, London, W.C., 12 Dec., 1885.
W. Carpenter, Esq., Baltimore.

Dear Sir,—I am obliged to you for the copy of your "One Hundred Proofs that the Earth is not a Globe," and for the evident kindness of your intention in dedicating the work to me. The only further remark it occurs to me to offer is that I call myself rather a student of astronomy than an astronomer.
Yours faithfully,
RICHARD A. PROCTOR.
P. S. Perhaps the pamphlet might more precisely be called "One hundred difficulties for young students of astronomy."

[Appendix to Fourth Edition.]
COPY OF LETTER FROM SPENCER F. BAIRD, ESQ.
Smithsonian Institution, Washington, D. C., Jan. 6, 1886.

Dear Sir,—A copy of your "One Hundred Proofs that the Earth is not a globe" was duly received, and was deposited in Library of Congress October 8, 1884. [1885] A pressure of much more important work has prevented any attempt at reviewing these hundred proofs:—which however have doubtless been thoroughly investigated by the inquisitive astronomers and geodesists of the last four centuries.
Yours very respectfully,
SPENCER F. BAIRD, Secretary S. I.
Mr. William Carpenter, 71, Chew Street, Baltimore, Md.

Copy of a letter from one of the several applicants for the "One Hundred Proofs" for the purpose of reviewing them. The writer is Professor of Mathematics at the High School, Auburn, N. Y., and, in his application for the pamphlet, says: "Am a Yale graduate and a Yale Law School man: took the John A Porter Prize (literary) ($250) at Yale College."

Auburn, Dec. 10th, 1885. My Dear Sir: Your treatise was received. I have looked it over and noted it somewhat. A review of it to do it justice would be a somewhat long and laborious task. Before I undertook so much thought I would write and ask What and how much you expect: how elaborately you wished it discussed: and what remuneration might be expected. It sets forth many new and strange doctrines which would have to be thoroughly discussed and mastered before reviewed. I am hard at work at present but would like to tackle this if it would be for my interest as well as yours. Hope you will let me know very soon. Very respectfully,
To Mr. W. Carpenter, Baltimore, Md. FRANK STRONG.

NOTE.—Unless a man be willing to sell his soul for his supposed worldly "interest," he will not dare to "tackle" the "One Hundred Proofs that the Earth is Not a Globe." No man with well-balanced faculties will thus condemn himself. We charge the mathematicians of the world that, if they cannot say what they think of this pamphlet in a dozen words, they are entitled to no other name than—cowards!
Baltimore, Maryland, May 22, 1886.

APPENDIX TO THE FIFTH EDITION.

Editorial from the "New York World," of August 2, 1886:—
THE EARTH IS FLAT.

The iconoclastic tendencies of the age have received new impetus from Mr. WILLIAM CARPENTER, who comes forward with one hundred proofs that the earth is not a globe. It will be a sad shock to many conservatives who have since their childhood fondly held to the conviction that "the earth is round like an orange, a little flattened at the poles." To find that, after all, we have been living all these years on a prosaic and unromantic plane is far from satisfactory. We have rather gloried in the belief that the semi-barbarous nations on the other side of the earth did not carry their heads in the same direction in which ours point. It is hard to accept the assertion that the cannibals on savage islands are walking about on the same level with the civilized nations of our little world.

But Mr. CARPENTER has one hundred proofs that such is the unsatisfactory truth. Not only that, but the iconoclast claims that we are not whirling through space at a terrible rate, but are absolutely stationary. Some probability is given to this proposition by the present hot weather. The earth seems to be becalmed. If it were moving at the rate of nineteen miles a second wouldn't there be a breeze? This question is thrown out as perhaps offering the one hundred and first proof that the earth is not a globe. Mr. CARPENTER may obtain the proof in detail at the office at our usual rates. A revolution will, of course, take place in the school geographies as soon as Mr. CARPENTER'S theories have been closely studied. No longer will the little boy answer the question as to the shape of the earth by the answer which has come ringing down the ages, "It's round like a ball, sir." No. He'll have to use the unpoetic formula, "It's flat like a pancake, sir."

But, perhaps, after we have become used to the new idea it will not be unpleasant. The ancients flourished in the belief that the earth was a great plane. Why shouldn't we be equally fortunate? It may be romantic but it is not especially comforting to think that the earth is rushing through space twisting and curving like a gigantic ball delivered from the hand of an enormous pitcher. Something in the universe might make a base hit if we kept on and we would be knocked over an aerial fence and never found. Perhaps, after all, it is safer to live on Mr. CARPENTER'S stationary plane.

The "Record," of Philadelphia, June 5, 1886, has the following, in the Literary Notes:—"Under the title One Hundred Proofs that the Earth is Not a Globe, Mr. William Carpenter, of Baltimore, publishes a pamphlet which is interesting on account of the originality of the views advanced, and, from his standpoint, the very logical manner in which he seeks to establish their truth. Mr. Carpenter is a disciple of what is called the Zetetic school of philosophy, and was referee for Mr. John Hampden when that gentleman, in 1870, made a wager with Mr. Alfred R. Wallace, of England, that the surface of standing water is always level, and therefore that the earth is flat. Since then he has combated his views with much earnestness, both in writing and on the platform, and, whatever opinions we may have on the subject, a perusal of his little book will prove interesting and afford room for careful study."

"The motto which he puts on the cover—'Upright, Downright, Straightforward'—is well chosen, for it is an upright lie, a downright invention, and a straightforward butt of a bull at a locomotive."—The Florida Times Union, Dec. 13, 1885. Editor, Charles H. Jones. [Pray, Mr. Jones, tell us what you mean by "an upright lie." ! !]

"We have received a pamphlet from a gentleman who thinks to prove that the earth is flat, but who succeeds only in showing that he is himself one."—New York Herald, Dec. 19, 1885. [The reviewer, in this case, is, no doubt, a very "sharp" man, but his honesty—if he have any at all—is jagged and worn out. The "quotations" which he gives are fraudulent, there being nothing like them in the pamphlet.]

"The author of the pamphlet is no 'flat,' though he may perhaps be called a 'crank.' "—St. Catharines (Can.) Evening Jour., Dec. 23.

"To say that the contents of the book are erudite and entertaining does not do Mr. Carpenter's astronomical ability half credit."—The Sunday Truth, Buffalo, Dec. 27, 1885.

"The entire work is very ingeniously gotten up. . The matter of perspective is treated in a very clever manner, and the coming up of 'hull-down' vessels on the horizon is illustrated by several well-worded examples."—Buffalo Times, Dec. 28, 1885.

"The erudite author, who travels armed with plans and specifications to fire at the skeptical at a moment's notice, feels that he is doing a good work, and that his hundred anti-globular conclusions must certainly knock the general belief in territorial rotundity out of time." . . "We trust that the distinguished author who has failed to coax Richard Proctor into a public discussion may find as many citizens willing to invest two shillings in his peculiar literature as he deserves."—Buffalo Courier, Dec. 27, 1885, and Jan. 1, 1886.

"It is a pleasure now to see a man of Mr. Carpenter's attainments fall into line and take up the cudgels against the theories of the scientists who have taught this pernicious doctrine [the sphericity of the earth]."—Rochester Morning Herald, Jan. 13, 1886.

"As the game stands now, there is 'one horse' for Prof. Carpenter." —Buffalo World, Jan. 16, 1886.

"It is interesting to show how much can be said in favor of the flat world theory. . It is fairly well written, although, we believe filled with misstatements of facts."—Rochester Democrat and Chronicle, Jan. 17, 1886. [We "believe" the editor cannot point one out.]

"It is certainly worth twice the price, and will be read by all with peculiar interest."—Scranton Truth, March 8, 1886.

"Mr. WILLIAM CARPENTER has come to Washington with a "hundred proofs that the earth is not a globe." He has a pamphlet on the subject which is ingenious, to say the least, and he is ominously eager to discuss the matter with any one who still clings to the absurd prejudices of the astronomers."—The Hatchet, May 9, 1886.

"It contains some curious problems for solution, and the author boldly asserts that until they are solved the globular theory of the earth remains unproven, and is fallacious, &c."—The Presbyterian, Philadelphia, June 19, 1886.

"His reasoning is, to say the least, plausible, and the book interesting."—The Item, Philadelphia, June 10, 1886.

"Mr. Carpenter seems to have made a thorough investigation of the subject, and his arguments are practical and to the point."—Sunday Mercury, Philadelphia, June 13, 1886.

"A gentleman has just called at the editorial rooms with a pamphlet which is designed to demonstrate that the earth is not a globe, but a flat disk; he also laid before us a chart from which it plainly appeared that the earth is a circular expanse of land, with the north pole in the exact center, and the Antarctic Sea flowing all around the land. . . We went on to state that we lodged the care of all astronomical questions in the hands of Rev. R. M. Luther, to whom these perplexing matters are but as child's play. . . . Our readers may, therefore, expect at an early date a judicial view of the astronomical and cosmological situation."—National Baptist, Philadelphia, July 8, 1886. Editor, Dr. Wayland. [We hope that the Rev. R. M. Luther will give us the means of publishing his decision before many more editions of the "Hundred Proofs" be issued. We are afraid that he finds the business much more than "child's play."]

"One Hundred Proofs that the Earth is Not a Globe," by William Carpenter, is published by the author, whose novel and rather startling position is certainly fortified by a number of argumentative points, which, if they do not shake the reader's preconceived notions on the subject, will, at least, be found entertaining for the style in which they are put."—Evening Star, Philadelphia, July 22, 1886.

"His 'Proofs' go a long way towards convincing many that his ideas on the subject are practical and sensible."—Fashion Journal, Philadelphia, July, 1886. Editor, Mrs. F. E. Benedict.

"'One Hundred Proofs that the Earth is Not a Globe' is a curious little pamphlet that we can commend to all interested in astronomy and related sciences. It may not upset received notions on the subject, but will give cause for much serious reflection. Published by the author, Wm. Carpenter, Baltimore, Md. Price 25 cents."—The Saturday Evening Post, Philadelphia, July 31, 1886.

"Here now is an able thinker of Baltimore, Professor WILLIAM CARPENTER, who presents the claims of the Zetetic philosophy to be considered the leading issue of our times. . One of the great proofs of the truth of the philosophy is that the regular astronomers do not dare to gainsay it. . . They are well aware there is no South pole. . . Prof. CARPENTER, in a treatise that has reached us, furnishes 100 proofs that the earth is flat, and while we cannot say that we understand all of them we appreciate the earnestness of his appeals to the moral people of the community to rise up and overthrow the miserable system of error that is being forced upon our children in the public schools, vitiating the very foundations of knowledge. What issue can be more noble or inspiring than Truth vs. Error? Here is an issue on which there can be no trifling or compromise. In the great contest between those who hold the earth is flat and they who contend that it is round, let the flats assert themselves."—Milwaukee Sentinel, Aug., 1886. [From a long article, "The Great Zetetic Issue."]

LETTERS TO PROFESSOR GILMAN, OF THE JOHNS HOPKINS UNIVERSITY.

71 Chew Street, Baltimore, September 10, 1886.

Prof. Gilman, Johns Hopkins University—Sir: On the 21st ultimo I wrote to ask you if you received the pamphlet, which I left for you at the University twelve months ago, entitled "One Hundred Proofs that the Earth is Not a Globe," and, if so, that you would kindly give me your opinion concerning it. I write, now, to ask you if you received my letter. I am quite sure that you will consider that the importance of the subject fully warrants the endeavor on my part to gain the views which may be entertained by you respecting it. The fifth edition will soon be called for, and anything you may urge —for or against—I shall be happy to insert in the "appendix." I send, herewith, a copy of the fourth edition of the pamphlet.

Yours sincerely, William Carpenter.

71 Chew Street, Baltimore, October 7, 1886.

Professor Gilman—Dear Sir: I am now preparing the appendix for the fifth edition of my "One Hundred Proofs that the Earth is Not a Globe," and I should be glad to receive your opinion of this work to insert in the said appendix. I can offer you from a few lines to a page, or two if necessary. Of course, if this work as a whole be a fraud, it must be fraudulent in all its parts; and each one of the "hundred proofs" must contain a fallacy of some kind or other, and the thing would justify your disapprobation—expressed in few words or many. If, on the other hand, the work is what it professes to be, it will certainly claim your approval. Yours sincerely, W. Carpenter.

71 Chew Street, Baltimore, October 14, 1886.

Prof. Gilman—Dear Sir: A week ago I wrote you a letter to tell you that I should be glad to receive your opinion of the "Hundred Proofs that the Earth is Not a Globe," of which work 5,000 copies are now in circulation. I wrote this work (26 pages) in one week, without neglecting my daily business: surely, you can reply to it in a week from this time. I will give you from one to four pages, if you wish that amount of space, and send you fifty copies, if you desire to have them, without putting you to the slightest expense. I will even take any suggestion you please to make as to the title which shall be given to this extra edition of my work containing your reply or opinions. I should be sorry to be under the necessity of printing this letter, with others, in my next edition, in the place of any such reply or expression of opinion; for I feel sure there is no one in Baltimore who is more capable of giving an opinion on this great subject. Trusting to hear from you in a few days, I am, Dear Sir, Yours truly,

William Carpenter.

71 Chew Street, Baltimore, October 22, 1886.

Prof. Gilman—Sir: This is the fifth letter—and the last—to you, asking you for an expression of your opinion concerning the "One Hundred Proofs that the Earth is Not a Globe." Which would you prefer—to see my words, or yours, in print? I give you a week in which to decide. Truly, William Carpenter.

THE JOHNS HOPKINS UNIVERSITY, OF BALTIMORE.

We are indebted to "Scribner's Monthly" for the following remarks concerning this institution:—"By the will of Johns Hopkins, a merchant of Baltimore, the sum of $7,000,000 was devoted to the endowment of a University and a Hospital, $3,500,000 being devoted to each. This is the largest single endowment ever made to an institution of learning in this country. To the bequest no burdensome conditions were attached." . . "The Physiological Laboratory of the Johns Hopkins has no peer in this country, and the other laboratories few equals and no superiors."

In the First Annual Report of the University (1876) we read:— "Early in the month of February, 1874, the Trustees of the University having been apprised by the Executors of Johns Hopkins, of the endowment provided by his will, took proper steps for organization and entering upon the practical duties of the trust, and addressed themselves to the selection of a President of the University. With this view the Trustees sought the counsel and advice of the heads of several of the leading seats of learning in the country, and, upon unanimous recommendation and endorsement from these sources, the choice fell upon Mr. DANIEL C. GILMAN, who, at the time, occupied the position of President of the University of California.

"Mr. Gilman is a graduate of Yale College, and for several years before his call to California, was a Professor in that institution, taking an active part in the organization and development of 'The Sheffield Scientific School of Yale College,' at New Haven. Upon receiving an invitation to Baltimore, he resigned the office which he had held in California since 1872, and entered upon the service of The Johns Hopkins University, May 1, 1875."—GALLOWAY CHESTON.

"In the hunt for truth, we are not first hunters, and then men; we are first and always men, then hunters."—D. C. GILMAN, Oct., 1883.

The "One Hundred Proofs that the Earth is Not a Globe" have been running around within the observation of the master huntsman and his men for a year or more: now let the hunters prove themselves to be men; and the men, hunters. It is impossible to be successful hunters for Truth, if Error be allowed to go scot-free. Nay, it is utterly impossible for the Johns Hopkins University to answer the purpose of its founder if its hunters for Truth do not first hunt Error with their hounds and hold it up to ridicule, and then, and always, keep a watchful eye for the Truth lest they should injure it by their hot haste or wound it with their weapons. Prof. DANIEL C. GILMAN, we charge you that the duties of your office render it imperative that, sooner or later, you lead your men into the field against the hundred proofs, to show the world that they are hunters worthy of the name— if, in your superior judgment, you decide that there is Error to be slain —or, show that your hunters are worthy of the better name of men, by inducing them to follow and sustain you, out of the beaten track, in your endeavors to uphold God's Truth, if, in your superior judgment, you tell them, "There is a Truth to be upheld!"

[End of the Appendix to the Fifth Edition. Nov. 9, 1886.]

CPSIA information can be obtained
at www.ICGtesting.com
Printed in the USA
BVOW09s1439241017
498539BV00010B/126/P